# Biennial Flight Review

- FAR's EXPLAINED
- AIRSPACE REVIEW
- FLIGHT MANEUVERS
- AIRPORT OPERATIONS
- ATC COMMUNICATIONS

PUBLISHED BY
FLIGHT TIME PUBLISHING
www.flight-time.com
BY ART PARMA CFI, ATP, A&P/IA

# CONTENTS

**FAR 61.56**

a) Except as provided in paragraphs (b) and (f) of this section, a flight review consists of a minimum of 1 hour of flight training and 1 hour of ground training. The review must include:

(1) A review of the current general operating and flight rules of part 91 of this chapter; and

(2) A review of those maneuvers and procedures that, at the discretion of the person giving the review, are necessary for the pilot to demonstrate the safe exercise of the privileges of the pilot certificate.

b) Glider pilots may substitute a minimum of three instructional flights in a glider, each of which includes a flight to traffic pattern altitude, in lieu of the 1 hour of flight training required in paragraph (a) of this section.

c) Except as provided in paragraphs (d), (e), and (g) of this section, no person may act as pilot in command of an aircraft unless, since the beginning of the 24th calendar month before the month in which that pilot acts as pilot in command, that person has -

(1) Accomplished a flight review given in an aircraft for which that pilot is rated by an authorized instructor; and

(2) A logbook endorsed from an authorized instructor who gave the review certifying that the person has satisfactorily completed the review.

d) A person who has, within the period specified in paragraph (c) of this section, passed a pilot proficiency check conducted by an examiner, an approved pilot check airman, or a U.S. Armed Force, for a pilot certificate, rating, or operating privilege need not accomplish the flight review required by this section.

e) A person who has, within the period specified in paragraph (c) of this section, satisfactorily

**FAR 61.56 (continued)**

accomplished one or more phases of an FAA-sponsored pilot proficiency award program need not accomplish the flight review required by this section.

f.  A person who holds a current flight instructor certificate who has, within the period specified in paragraph (c) of this section, satisfactorily completed a renewal of a flight instructor certificate under the provisions in § 61.197 need not accomplish the 1 hour of ground training specified in paragraph (a) of this section.

g.  A student pilot need not accomplish the flight review required by this section provided the student pilot is undergoing training for a certificate and has a current solo flight endorsement as required under § 61.87 of this part.

h.  The requirements of this section may be accomplished in combination with the requirements of § 61.57 and other applicable recent experience requirements at the discretion of the authorized instructor conducting the flight review.

i.  A flight simulator or flight training device may be used to meet the flight review requirements of this section subject to the following conditions:

(1) The flight simulator or flight training device must be used in accordance with an approved course conducted by a training center certificated under part 142 of this chapter.

(2) Unless the flight review is undertaken in a flight simulator that is approved for landings, the applicant must meet the takeoff and landing requirements of § 61.57(a) or § 61.57(b) of this part.

(3) The flight simulator or flight training device used must represent an aircraft or set of aircraft for which the pilot is rated.

1. **Who is required to have a Biennial Flight Review?**
Every pilot needs to complete a flight review within the preceding 24 calendar months to act as pilot in command whether flying solo or carrying passengers. If a pilot completes a BFR December 8, 2000, he/she may act as pilot in command until December 31, 2002.

2. **Is a written examination required?**
No. The pilot is required to demonstrate his/her knowledge of the general operating and flight rules of FAR Part 91.

3. **What counts as a flight review?**
The issuance of a new certificate, rating or operating privilege, or the completion of any phase of the FAA "Wings" program will count as a flight review. An Instrument Proficiency Check (IPC) does not qualify as a flight review but may be combined in the same flight.

4. **Is it possible to fail a flight review?**
A person cannot "fail" a flight review however it is the discretion of the instructor whether or not to sign the pilot's logbook as having completed the review satisfactorily.

5. **If my flight review is unsuccessful will it be recorded in my pilot logbook or is any notice forwarded to the FAA?**
No, and no notice is forwarded to the FAA.

6. **Is my pilot's certificate still valid if I do not have a current flight review?**
Yes. Your pilot certificate is valid until surrendered, suspended or revoked. You may not act as pilot-in-command until you have completed a flight review however, you may fly as a safety pilot as long as you agree that you will not be the acting PIC.

**7. Is a flight review required to fly solo?**
Yes. A pilot is acting as pilot-in-command when flying solo and must have a flight review.

**8. Must I have a current medical certificate to complete a flight review?**
You may complete a flight review without a current medical certificate but you must obtain a current medical certificate before acting as pilot in command. The instructor may act as pilot in command during the flight review.

**9. Do I have to take a flight review for each category and class aircraft in which I am rated?**
No. Completion of a flight review allows you to act as pilot in command in any category and class aircraft for which you are rated.

**10. How much time is required for a flight review?**
A minimum of 1 hour flight instruction and 1 hour ground instruction is required.

**11. I have an instrument rating. Will my flight review include instrument maneuvers and approaches?**
Not necessarily. You must demonstrate those maneuvers and procedures determined by the flight instructor as necessary to demonstrate you can safely exercise the privileges of your certificate.

**12. What logbook endorsement is required?**
Completion of a BFR must be endorsed in the pilot's logbook and should read substantively as follows:

Mr./Mrs. _________________ holder of pilot certificate
#_________________ has satisfactorily completed the review required by FAR 61.56(a) on 12-08-2006.
s/s J. Jones 123456CFI, expires 10-31-2007.

## PILOT CURRENCY REQUIREMENTS

- **Medical Certificate**     **60/24 Months**
- **Biennial Flight Review**  **24 Months**
- **Recent Experience**       **90 Days**

1. **Medical Certificate.** Within the preceding 60 calendar months (age under 40) or 24 months (age over 40), for private pilot privileges. (FAR 61.23d, FAR 61.3c)
2. **Flight Review.** Within the preceding 24 calendar months. (FAR 61.56)
3. **Day Recent Flight Experience.** To carry passengers within the preceding 90 days, 3 takeoffs and landings as the sole manipulator of the flight controls, in the same category and class.
   *Three take-offs and landings in a single-engine airplane will not qualify a pilot to carry passengers in a multi-engine airplane. (FAR 6l.57a)*
   **Tailwheel Aircraft.** If the aircraft is a tailwheel airplane the landings must be to a full stop in a tailwheel airplane. (FAR 61.57a)
   **Night Recent Flight Experience.** To carry passengers at night, three takeoffs and landings to a full stop within the period from one hour after sunset to one hour before sunrise in the same category and class within the preceding 90 days.
   *The night experience requirement refers to category and class but does not mention tailwheel airplanes. Therefore if you do 3 daytime stop and goes in a Piper Cub in the preceding 90 days and 3 nighttime stop and goes in a Cessna 152 in the same period, you can legally carry a passenger at night in a Piper Cub. (FAR 61.57b)*

## AIRCRAFT CURRENCY REQUIREMENTS

- **Annual Inspection**     **12 Months**
- **Transponder/Altimeter**     **24 Months**
- **ELT/Battery**     **12/24 Months**
- **100 Hour Inspection**     **Aircraft for Hire**

1. **Annual Inspection.** Within the preceding 12 calendar months. *The annual inspection is entered in the aircraft maintenance logs and should note the next compliance due of any recurring Airworthiness Directives. (FAR 91.409)*
2. **Transponder/Altimeter System Test.** Within the preceding 24 calendar months. (FAR 91.411, 91.413)
3. **ELT Battery Replacement:** Within 24 months or 1 hour cumulative use. Expiration date marked on transmitter and in aircraft maintenance record.
   **ELT Operational check:** Within 12 calendar months and noted in aircraft maintenance record. (FAR 91.207)
4. **100 Hour Inspection.** Within the preceding 100 hours an annual or 100 hour inspection if the aircraft is operated for hire. (FAR 91.409)

## AIRCRAFT DOCUMENTS REQUIRED (A.R.R.O.W.)

- **A**irworthiness Certificate
- **R**egistration Certificate
- **R**adio Station License (International Flights)
- **O**perating Limitations
- **W**eight and Balance Information

## AIRCRAFT AIRWORTHINESS FAR 91.7

1. The owner or operator of the aircraft is responsible for maintaining the aircraft in an airworthy condition.
2. The pilot in command is responsible for determining the aircraft is in condition for a safe flight.

## PREFLIGHT ACTION REQUIRED 91.103

### For any flight:

1. The runway lengths of intended use.
2. Takeoff and landing performance of the aircraft.

### For a flight not in the vicinity of the airport or an IFR flight:

1. Weather reports and forecasts.
2. Fuel requirements.
3. Alternatives available if the flight cannot be completed as planned.
4. Any known traffic delays advised by ATC.

## PILOT/CO-PILOT USE OF SAFETY BELTS 91.105

1. A required crewmember (pilot) must wear a safety belt and shoulder harness during takeoff and landing.
2. A required crewmember (pilot) must keep a safety belt fastened while seated at his station enroute.

## PASSENGER USE OF SAFETY BELTS 91.107

### Before taxi, takeoff or landing an aircraft:

1. The pilot must ensure each person has been briefed on how to fasten and unfasten his or her safety belt.
2. Before takeoff or landing the pilot must notify each occupant to fasten his or her safety belt.
3. Each person must occupy an approved seat with a safety belt, and if installed, a shoulder harness, properly secured.
4. Child car seats used in airplanes should have a label stating, "THIS RESTRAINT IS APPROVED FOR USE IN MOTOR VEHICLES AND AIRCRAFT". The car seat should be secured in a forward facing seat and a person must be designated to attend to the safety of the child during the flight.
   *(A child under the age of two may still be held by an adult in a seat however a child car seat should be used whenever possible).*

# INOPERATIVE INSTRUMENTS AND EQUIPMENT  91.213

1.   No person may take off an aircraft unless the inoperative equipment is deactivated and marked "Inoperative" and it is not part of the minimum equipment required for day or night VFR or IFR flight as applicable or required in the equipment list by certification of the aircraft.

# INSTRUMENTS REQUIRED FOR VFR DAY 91.205

Gas gauge.
Oil pressure gauge.
Oil temperature gauge.
Seat belts and shoulder straps.
Emergency locator transmitter.

Altimeter.

Compass.
Airspeed indicator.
Tachometer.

# INSTRUMENTS REQUIRED FOR VFR NIGHT 91.205

Anti-collision light system.
Position lights.
Energy source.
Spare fuses. One complete set or 3 of each kind.

# ALCOHOL OR DRUGS 91.17

**No person may operate an aircraft:**

1.   Within 8 hours after consuming an alcoholic beverage.
2.   While under the influence of alcohol.
3.   While using any drug that may affect the persons ability to fly safely.
4.   While having .04 percent by weight or more alcohol in the blood.
5.   A pilot may not carry a passenger that appears to be intoxicated or under the influence of drugs.

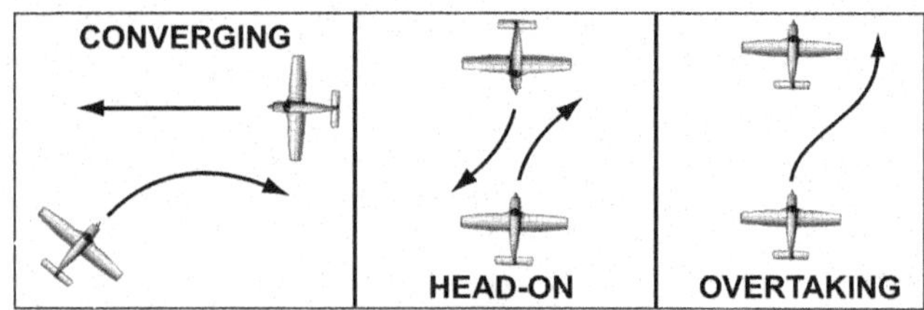


## BALLOON > GLIDER > AIRSHIP >ALL POWERED AIRCRAFT
(Converging aircraft, same altitude, of different categories)

1. **In distress.** An  aircraft in distress has the right of way over all other air traffic.
2. **Converging.** When aircraft of the same category are converging at the same altitude (except head-on or nearly so), the aircraft to the others right has the right of way. If the aircraft are of different categories:
   a)  A balloon has the right-of-way over any other category.
   b)  A glider has the right-of-way over an airship, airplane or rotorcraft.
   c)  An airship has the right-of-way over an airplane or rotorcraft.
   d)  Towing or refueling aircraft has the right-of-way over all other engine driven aircraft.
3. **Approaching head-on.** Aircraft approaching head-on shall alter course to the right.
4. **Overtaking.** Aircraft being overtaken have the right-of-way and the overtaking aircraft shall alter course to the right to pass well clear.
5. **Landing.** Aircraft on final approach or while landing have the right-of-way over other aircraft in flight or on

# RIGHT OF WAY RULES 91.113 (continued)

the ground but shall not take advantage of this rule to force another aircraft off the runway that has already landed and is attempting to clear the runway. When two or more aircraft are approaching to land the aircraft at the lower altitude has the right-of-way, but it should not take advantage of this rule to overtake or cut in front of another aircraft.

# AIRCRAFT SPEED LIMITATIONS 91.117

* Below 10,000' MSL: **250 KIAS (288 MPH)**

* At or below 2,500' AGL within 4 NM of a primary airport of a Class C or D airspace: **200 KIAS (230 MPH)**

* In a Class B airspace VFR corridor* or underneath Class B airspace: **200 KIAS (230 MPH)**
*Or as noted on chart.

# MINIMUM SAFE ALTITUDES 91.119

Except when necessary for takeoff or landing, no person may operate an aircraft below the following altitudes:

1. **Anywhere** - An altitude that allows an emergency landing without undue hazard to persons and property on the ground in the event of a power failure.
2. **Congested areas** - 1,000' above the highest obstacle within a 2,000' radius of the aircraft.
3. **Other than congested areas** - 500' above the surface or no closer than 500' to any person, vessel, vehicle or structure.

## FLIGHT OPERATIONS OPERATING NEAR OTHER AIRCRAFT 91.111

1.  No pilot may operate an aircraft close to another aircraft as to create a collision hazard.
2.  Aircraft may not fly in formation flight unless agreed to by the pilot in command of each aircraft.
3.  Aircraft carrying passengers for hire may not fly in formation flight.

## ATC LIGHT SIGNALS 91.125

| ATC LIGHT SIGNALS FAR 91.125 | | |
|---|---|---|
| COLOR AND TYPE OF SIGNAL | ON THE GROUND | IN FLIGHT |
| STEADY GREEN | CLEARED FOR TAKEOFF | CLEARED TO LAND |
| FLASHING GREEN | CLEARED TO TAXI | RETURN FOR LANDING (TO BE FOLLOWED BY A STEADY GREEN) |
| STEADY RED | STOP | GIVE WAY TO OTHER AIRCRAFT AND CONTINUE CIRCLING |
| FLASHING RED | TAXI CLEAR OF RUNWAY IN USE | AIRPORT UNSAFE - DO NOT LAND |
| FLASHING WHITE | RETURN TO STARTING POINT ON THE AIRPORT | NOT APPLICABLE |
| ALTERNATING RED AND GREEN | GENERAL WARNING - USE EXTREME CAUTION | |

## COMMUNICATIONS FAILURE PROCEDURES

* Squawk **7600** on the transponder.
* Stay clear until the direction of traffic in the pattern can be determined.
* Tune receiver or transmit on the tower frequency as able.
* Flash your landing light or wave your wings when able to acknowledge transmissions or light signals.

# FUEL REQUIREMENTS VFR DAY 91.151

No person may begin a flight in an airplane unless (considering wind and forecast weather) there is enough fuel to fly to the point of intended landing and:

- **DAY**     Destination + 30 minutes at normal cruise
- **NIGHT**   Destination + 45 minutes at normal cruise

*Note: Pilots should realize these minimum requirement although legal may be inadequate, especially at night and in poor weather.*

## EMERGENCY LOCATOR TRANSMITTER 91.207

- Transmits on 121.5 MHz.
- Not required on training flights within a 50 mile radius.
- Battery replaced every 2 years or after 1 hour use.
- Operational check every 12 months.

## AIRCRAFT LIGHTS 91.209

No person may operate, park or move an aircraft unless the aircraft has:

- Lighted position lights from sunset to sunrise.
- Lighted anticollision lights (if equipped), day or night, unless the pilot determines the lights should be turned off in the interest of safety.

## AUTOMATIC DEPENDENT SURVEILLANCE-BROADCAST (ADS-B)  91.225

- Above 10,000' MSL.
- Within Class A, B, and C airspace.
- Within 30 nautical radius of Class B primary airport.
- Over-flying Class B or Class C airspace.
- ADS-B operate and transmit in controlled airspace.

| TRANSPONDER OPERATION ||
|---|---|
| 1200 | • VFR |
| 7500 | • Hijack |
| 7600 | • Lost Communications |
| 7700 | • Emergency |

# VFR WEATHER MINIMUMS 91.155

| VFR WEATHER MINIMUMS FAR 91.155 | | |
|---|---|---|
| **Airspace** | **Flight Visibility** | **Distance From Clouds** |
| **CLASS A** | Not Applicable | Not Applicable |
| **CLASS B** | 3 Statute Miles | Clear of Clouds |
| **CLASS C**<br>**CLASS D**<br>**CLASS E** | 3 Statute Miles | 500' below<br>1,000' above<br>2,000' horizontal |
| **CLASS E & G**<br>At and above<br>10,000' MSL | 5 Statute Miles | 1,000' below<br>1,000' above<br>1 mile horizontal |
| **CLASS G DAY**<br>Above 1,200' AGL | 1 Statute Mile | 500' below<br>1,000' above<br>2,000' horizontal |
| **CLASS G NIGHT**<br>Above 1,200' AGL | 3 Statute Miles | 500' below<br>1,000' above<br>2,000' horizontal |
| **CLASS G DAY**<br>Below 1,200' AGL | 1 Statute Mile | Clear of Clouds |
| **CLASS G NIGHT**<br>Below 1,200' AGL | 3 Statute Miles | 500' below<br>1,000' above<br>2,000' horizontal |

# SUPPLEMENTAL OXYGEN 91.211

ALL OCCUPANTS MUST BE SUPPLIED WITH SUPPLEMENTAL OXYGEN

**15,000' MSL**

MINUMUM FLIGHT CREW MUST USE
OXYGEN FOR THE ENTIRE FLIGHT TIME

**14,000' MSL**

MINUMUM FLIGHT CREW MUST USE OXYGEN
AFTER 30 MINUTES OF FLIGHT TIME

**12,500' MSL**

NO SUPPLEMENTAL OXYGEN REQUIRED

*Oxygen recommended above 8,000' MSL at night.

## VFR CRUISING ALTITUDES 91.159

Aircraft flying at more than 3,000' AGL in level cruising flight shall maintain the appropriate VFR cruising altitude.

**Magnetic Course
(0° through 179°)**

**ODD altitudes + 500'
(3,500', 5,500' MSL etc.)**

**Magnetic Course
(180° through 359°)**

**EVEN altitudes + 500'
(4,500', 6,500' MSL etc.)**

## AEROBATIC FLIGHT 91.303

**No person may operate an aircraft in aerobatic flight:**

1. Over a congested area of a city, town or settlement.
2. Over an open assembly of persons.
3. Within the lateral boundaries of the surface areas of Class B, Class C, Class D or Class E airspace designated for an airport.
4. Within 4 nautical miles of the center line of any Federal airway.
5. Below an altitude of 1,500 feet above the surface.
6. With a flight visibility of less than 3 miles.

*For the purposes of this section, aerobatic flight means an intentional maneuver involving an abrupt change in an aircraft's attitude, an abnormal attitude, or abnormal acceleration, not necessary for normal flight*

## USE OF PARACHUTES 91.307(c)

Unless each occupant of the aircraft is wearing an approved parachute, no pilot shall execute any intentional maneuver that exceeds;

- A bank of 60° relative to the horizon;
- A nose up or nose down attitude of 30° to the horizon.

*The requirement for wearing parachutes does not apply while performing spins or any other flight maneuvers required by the regulations for any certificate or rating with a certified flight instructor.*

## UNCONTROLLED AIRPORTS
### Arriving Aircraft:

1. Familiarize yourself with the airport using an airport guide or the Airport Facilities Directory.
2. Monitor the CTAF 10-15 miles from the airport. Self-announce your position on the CTAF.
3. If necessary, over-fly the airport 500 feet above the traffic pattern altitude and look for the segmented circle and wind sock or landing direction indicator.
4. Descend to traffic pattern altitude at least 1/2 mile before entering the airport traffic pattern.
5. Fly a left hand traffic pattern unless otherwise indicated.
6. Make a standard 45° entry or an appropriate upwind, downwind, or crosswind entry. Do not make a straight-in approach.
7. Maintain traffic pattern altitude unless otherwise required by the applicable distance from cloud criteria (FAR 91.155). 1,000' AGL is standard.
8. Make the turn to final at least 1/4 mile from the approach end of the runway.
9. Self-announce your position on the CTAF at 10 miles from the airport, entering downwind, base, final and leaving the runway.

### Departing Aircraft:

1. Monitor the CTAF during taxi and run-up.
2. Look 360° for inbound aircraft not in radio contact before taxiing on to and taking off on the active runway.
3. Announce your takeoff and runway direction and your departure intentions.
4. Start the crosswind turn upon reaching 300' below the TPA to avoid climbing in the downwind leg.
5. Monitor the CTAF until 10 miles away from the airport.

### Local Airport Advisory
In Alaska **FSS** will provide advisory service to arriving and departing aircraft within a 10 mile radius of an airport where a **FSS** is located but there is not an operating control tower.

# TOWER CONTROLLED AIRPORTS

## Arriving Aircraft:

1. Listen to the ATIS information 15-20 miles out.
2. Contact the tower 10 to 15 miles from the airport.
3. VFR radar service is available at some airports to assist the controller in identifying aircraft and providing traffic advisories. The pilot remains solely responsible for collision avoidance and navigation.
4. Enter downwind left traffic unless otherwise instructed by the tower. (i.e.. straight-in approach, base entry, right traffic, etc.)
5. Notify the tower if you find it necessary to make major maneuvers for spacing. Always ask permission before making a 360° turn in the traffic pattern.
6. Make sure you have a clearance before landing.

## Departing Aircraft:

1. Monitor ATIS before contacting clearance delivery or ground control.
2. Taxi as instructed.
3. Repeat all hold short instructions given by the tower.
4. Remain on the tower frequency until outside the Class D airspace or released by the tower controller.

**UNICOM (Airports without an operating control tower) (122.7, 122.8, 123.0, 122.72, 122.97, 123.05, 123.07)**
May provide wind, weather, recommended runway, traffic pattern in use and any reported traffic. (122.95 is a UNICOM frequency used at airports with a control tower or FSS and may be used for fuel and transportation information. Wind direction and runway information may not be available on 122.95.)

**MULTICOM (122.9)**
Airports without a control tower, FSS or UNICOM. Pilots may share airport information and self-announce position and intentions. May be used for fuel and transportation information.

**Common Traffic Advisory Frequency (CTAF)**
May be UNICOM, MULTICOM, FSS, or tower frequency.

## CLASS A AIRSPACE 91.135

- IFR only, Instrument Rating required.
- Mode C transponder, ADS-B required.
- DME above 24,000' MSL  (FAR 91.205e)

Class A airspace extends from 18,000 feet MSL up to and including 60,000 feet MSL including the airspace overlying the water within 12 NM off the coast of the 48 contiguous states and Alaska.

## CLASS B AIRSPACE 91.131

- Receive an ATC clearance.
- Operational transponder with Mode C, ADS-B.
- Two-way radio communication capabilities.
- VOR receiver for IFR operations.
- VFR flight must remain clear of clouds.
- Private Pilot Certificate required for airports listed in Appendix D.
- Student, Sport and Recreational pilots with required flight and ground instruction and logbook endorsements.

Class B airspace is defined using VOR radials and DME arcs. Avoid flying close to the floor of overlying Class B airspace. Use the VFR Flyway Planning Chart on the back of the VFR Terminal Area Chart to avoid Class B airspace.

## CLASS C AIRSPACE 91.130

- Two-way radio.
- Establish communications before entering the airspace.
- Maintain communications while in Class C airspace.
- Operational transponder with Mode C, ADS-B.

The inner and outer circles (5 and 10 NM) extend from the primary airport of the Class C airspace. The outer area extends 20 nautical miles and is non-regulatory. *Communications are established when the controller repeats your call sign. You may enter the airspace unless you are asked to remain clear of the airspace.*

## CLASS D AIRSPACE 91.129

* Two-way radio or prior arrangement with ATC.
* Establish two-way communications prior to entering
  the airspace.

Class D airspace extends up to approximately 2,500
feet AGL and is depicted on the chart in hundreds of
feet MSL. When departing from an uncontrolled satellite
airport the pilot should establish communications with
ATC as soon as practicable after departure.

## CLASS E AIRSPACE (General Controlled Airspace)

* Transponder/Mode C on operation required if equipped.
* ATC clearance/flight plan required for IFR operations.

Class E airspace, is outlined on Sectional and Terminal
Area Charts with a magenta or blue vignette. The floor
of Class E airspace starts at 700 or 1,200 feet above the
surface or at designated MSL altitudes in mountainous
areas. Dashed magenta lines indicate where Class E
airspace is to the surface. Types of Class E airspace
include; Federal Airways, designated airport surface
areas and extensions for airports with an IAP, and
Offshore Airspace Areas and Enroute Domestic Areas
to provide IFR enroute services.

## CLASS G AIRSPACE (Uncontrolled Airspace)

* Transponder on operations not required*.
* IFR flight plan or clearance not required.

Class G (uncontrolled) airspace includes all the airspace
not designated as Class A, B, C, D, or E airspace.  Class
G airspace extends from the surface to 14,500 feet MSL
unless otherwise designated on chart.
*Note: It is recommended to operate the transponder
and Mode C function at all times if equipped.*

# SPECIAL USE AIRSPACE

**Prohibited Area** - Prohibited Areas are established in the interest of national safety. Do not enter prohibited airspace.

**Restricted Area** - Entering restricted airspace without prior permission may be extremely hazardous due to activities such as aerial gunnery and guided missiles

**Warning Area** - A Warning Area is legally international airspace (beyond the 3-mile limit offshore) and may not be charted as restricted airspace but may be equally as dangerous.

**Military Operations Area** - Prior permission is recommended but not required to enter a **MOA** however exercise caution and be vigilant for fast moving military aircraft. For current military activity contact **FSS**.

**Alert Area** - An Alert Area indicates an unusual type of aerial activity or high volume of pilot training. All aircraft both participating and nonparticipating are responsible for collision avoidance.

**Military Training Routes** - **MTR's** are charted on sectional charts as light gray lines with an associated **VR** or **IR** number (visual or instrument routes). Width of the routes may be several miles wide. **MTR's** indicate possible high speed military activity outside Military Operation Areas and are identified as follows:

- **4 digit numbers - Surface to 1,500' AGL**
- **3 digit numbers - Above 1,500' AGL**

Contact a **Flight Service Station (FSS)** to find out which routes are active. Exercise extreme vigilance when flying near Military Training Routes that are "hot".

# FLIGHT MANEUVERS

## NORMAL TAKEOFF

1. Flaps -- SET.
2. Magnetos -- BOTH.
3. Engine Instruments -- CHECK.
4. Strobes, Lights, Transponder -- ON.
5. Carburetor Heat -- OFF.
6. Throttle -- FULL OPEN.
7. Rotate -- _______ KTS.
8. Climb Out -- BEST RATE ($V_Y$) _______ KTS.

## CLIMB

1. Throttle -- FULL OPEN.
2. Mixture -- RICH (above 3,000' LEAN for max. RPM).
3. Airspeed -- NORMAL CLIMB _______ KTS.
4. Rudder -- CENTERED BALL.
5. Electric Fuel Pump -- OFF (at a safe altitude).

## STRAIGHT AND LEVEL FLIGHT

1. Power -- ADJUST to _______ RPM.
2. Trim -- ADJUST.
3. Mixture -- LEAN as required.
4. Engine Instruments -- CHECK.

## DESCENT

1. Carburetor Heat -- AS REQUIRED.
2. Power -- REDUCE to _______ RPM.
3. Mixture -- ADJUST as necessary.
4. Rudder -- CENTERED BALL.

## BEFORE LANDING

1.  Seat Belts, Shoulder Straps -- ADJUST and LOCK.
2.  Fuel Selector Valve -- ON, BOTH, or PROPER TANK.
3.  Electric Fuel Pump -- ON (if equipped).
4.  Landing Gear -- DOWN and LOCKED.
5.  Mixture -- FULL RICH.
6.  Brakes -- TEST.

## ABEAM TOUCHDOWN

1.  Carburetor Heat -- AS REQUIRED.
2.  Power -- REDUCE to __________ RPM.
3.  Flaps -- 10°.
4.  Airspeed -- __________ KTS.
5.  Altitude -- HOLD.
6.  Trim -- ADJUST.

## BASE LEG

1.  Power -- AS REQUIRED.
2.  Flaps -- __________ °.
3.  Airspeed -- __________ KTS.
4.  Trim -- ADJUST.

## FINAL APPROACH

1.  Power -- AS REQUIRED.
2.  Flaps -- FULL (when runway is assured).
3.  Airspeed -- __________ KTS (increase in gusty conditions).
4.  Trim -- ADJUST.

## LANDING

1.  Airspeed -- __________ KTS.
2.  Power -- IDLE when gliding touchdown is assured.
3.  Touchdown -- MAIN WHEELS FIRST.
4.  Braking -- MINIMUM REQUIRED. Slow before turning.

## AFTER LANDING

1.  Flaps -- RETRACT.
2.  Transponder -- STANDBY.
3.  Carburetor Heat -- OFF.

## TOUCH-AND-GO

1. Flaps -- RETRACT (verify).
2. Trim -- SET.
3. Carburetor Heat -- OFF.
4. Power -- FULL THROTTLE.

## GO-AROUND

**Objective:** To discontinue an approach for safety reasons i.e. another aircraft on the runway, unstabalized final approach, or wake turbulence avoidance.

1. Power -- FULL THROTTLE.
2. Carburetor Heat -- OFF.
3. Elevator -- STOP DESCENT.
4. Flaps -- RETRACT TO _________ °.
5. Airspeed -- _________ KTS.
6. Flaps -- RETRACT SLOWLY while climbing.

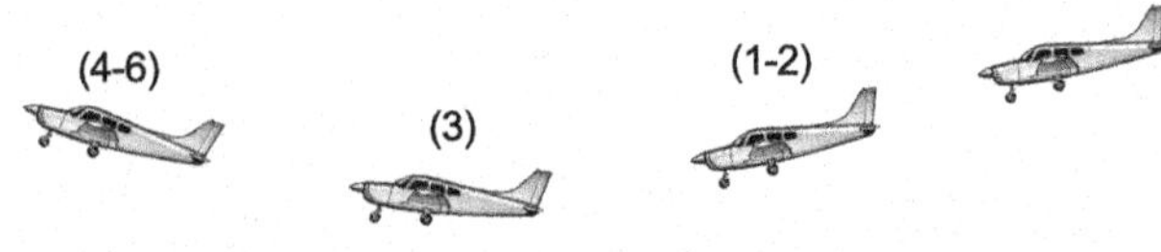


## TECHNIQUES

- Use all available power.
- Use the elevator to stop the descent.
- If flaps are full down, carefully retract the flaps one setting.
- Maintain a positive rate of climb and accelerate to $V_x$ while slowly retracting the flaps to 0°.
- Retract landing gear and open cowl flaps (if equipped) after establishing a positive rate of climb.
- Use caution, a fully loaded aircraft at a high density altitude *(high, hot, humid)* with full flaps, may not achieve a positive rate of climb.

# SHORT FIELD TAKEOFF

**Objective:** To take off from a short runway and climb as steeply as possible to clear simulated obstacles at the departure end of the runway.

1. Flaps -- AS REQUIRED.
2. Carburetor Heat -- OFF.
3. Mixture -- RICH (Lean above 3,000' for maximum RPM).
4. Brakes -- HOLD.
5. Throttle -- FULL OPEN smoothly and promptly.
6. Brakes -- RELEASE.
7. Elevator -- NEUTRAL or slightly tail low.
8. Rotate -- _________ KTS.
9. Airspeed -- _________ KTS until obstacles are cleared.
10. Flaps -- RETRACT SLOWLY while climbing.

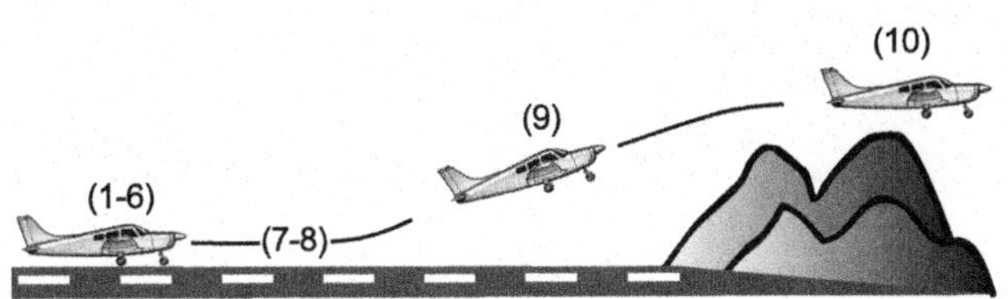


## TECHNIQUES

- Use the entire length of the runway .
- Hold the brakes while applying full power. Check for maximum engine performance before releasing brakes.
- Apply positive, prompt elevator back pressure to rotate approximately 5 knots before reaching $V_x$.
- Climb out at $V_x$ (+5,-0 knots).
- Lean the mixture at higher altitudes for best performance.

## SHORT FIELD LANDING

**Objective:** To land over a simulated or actual 50' obstacle using minimum approach airspeed, touch down at the beginning of the runway, and stop in the shortest possible distance.

1. Flaps -- FULL.
2. Airspeed -- _________ KTS.
3. Power -- SLIGHT AMOUNT to control descent.
4. Touchdown -- FIRST 100 feet of the runway.
5. Throttle -- CLOSE.
6. Brakes -- APPLY HEAVILY without skidding tires.
7. Flaps -- RETRACT.
8. Elevator -- FULL AFT while braking.

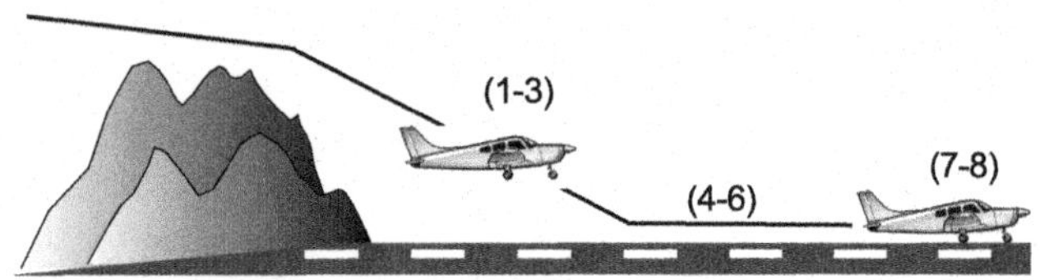


## TECHNIQUES

- Maintain a slight amount of power to control the rate of descent
- Allow the wheels to touch down as close as possible to the numbers. Use friction and braking to shorten ground roll.
- Raise the flaps and use back elevator pressure while braking to transfer as much weight as possible to the main wheels.

# SOFT FIELD TAKEOFF

**Objective:** To take off from a soft or rough surface and transfer the weight of the airplane from the wheels to the wings as soon as possible.

1.    Flaps -- AS REQUIRED.
2.    Brakes -- NONE.
3.    Throttle -- FULL OPEN smoothly and promptly.
4.    Elevator -- HOLD nose wheel off runway.
5.    Rudder -- MAINTAIN RUNWAY HEADING.
6.    Liftoff -- LOWER NOSE. Accelerate in ground effect.
7.    Airspeed -- CLIMB AT _________ KTS.
8.    Flaps -- RETRACT SLOWLY while climbing.

# TECHNIQUES

- More than normal right rudder will be required to maintain runway heading once the nose wheel is not in contact with the ground.
- Hold the elevator full aft while taxiing and at the beginning of the takeoff roll.
- Do not use the brakes while taxiing.

# SOFT FIELD LANDING

**Objective:** To land or simulate a landing on a soft surface such as wet grass or soft dirt. It is important to touch down smoothly at a minimum rate of descent and ground speed, and hold the nose wheel off the surface as long as possible.

1.　Flaps -- FULL.
2.　Airspeed -- _________ KTS.
3.　Power -- _________ RPM.
4.　Touchdown -- SMOOTHLY at minimum airspeed.
5.　Throttle -- CLOSE.
6.　Brakes -- NONE.
7.　Elevator -- HOLD FULL AFT.

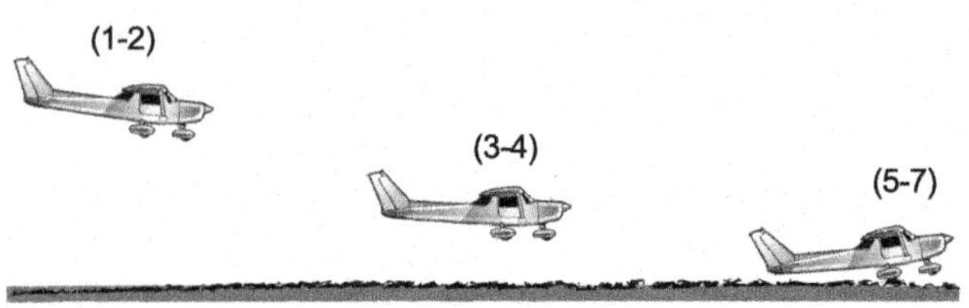


## TECHNIQUES

- Use the same approach configuration as a short field approach.
- Maintain the power setting throughout the flare until the wheels have made contact with the runway.
- Do not us the brakes during the rollout.
- Touch down smoothly and hold the nose wheel off of the runway as long as possible.

# CROSSWIND TAKEOFF

**Objective:** To maintain directional control of the airplane while on the runway and maintain a flight path along the extended runway centerline during climbout.

1. Ailerons -- FULL DEFLECTION INTO THE WIND.
2. Throttle -- APPLY FULL POWER.
3. Rudder -- MAINTAIN RUNWAY CENTERLINE.
4. Ailerons -- DECREASE DEFLECTION as necessary as airspeed increases.
5. Rotate -- $V_Y \pm 5$ KTS.
6. Climbout -- LEVEL WINGS. CRAB INTO THE WIND.

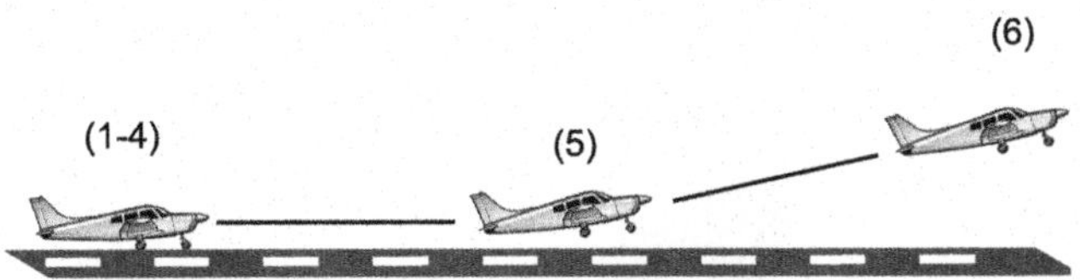


# TECHNIQUES

- Hold the plane on the runway longer than normal.
- Apply positive elevator back pressure to lift off immediately with little or no side skipping on the runway.
- Immediately after lift off point the nose of the airplane into the wind with the rudder pedals and level the wings with the ailerons. Do not climb out cross-controlled.
- Establish a crab angle to maintain runway centerline.

# CROSSWIND LANDINGS

**Objective:** To maintain the runway centerline using a side slip maneuver to compensate for the wind and to touch down on the runway without imposing side loads on the landing gear.

1. Flaps -- 0° to 25°.
2. Airspeed -- SLIGHTLY HIGHER THAN NORMAL.
3. Rudder -- OPPOSITE RUDDER to align the airplane fuselage with the runway.
4. Aileron -- LOWER UPWIND WING into the wind.
5. Flight Path -- MAINTAIN RUNWAY CENTERLINE USING AILERONS.
6. Heading -- KEEP THE FUSELAGE ALIGNED WITH THE RUNWAY USING RUDDER PRESSURES.
7. Touchdown -- UPWIND WHEEL FIRST.
8. Roll Out -- HOLD AILERON CORRECTIONS into the wind. Maintain directional control using rudder.

## TECHNIQUES

- Line up with the runway as soon as practical and determine the amount of rudder pressure required.
- During the flare as airspeed decreases more rudder and aileron deflections may be required.
- Hold the necessary correction throughout the landing flare.
- Use full aileron deflections as the airplane slows on the ground.

# TAKEOFF AND DEPARTURE STALLS

**Objective:** To demonstrate a full stall and recovery in the takeoff configuration with minimum loss of altitude.

1. Area -- CLEAR.
2. Carburetor Heat -- AS REQUIRED.
3. Power -- REDUCE to _______ RPM.
4. Airspeed -- SLOW to ROTATION SPEED ($V_R$) _____ KTS.
5. Carburetor Heat -- OFF.
6. Power -- FULL THROTTLE or takeoff climb power setting.
7. Rudder -- RIGHT RUDDER PRESSURE to center ball.
8. Elevator -- INCREASE BACK PRESSURE until stall occurs.

**Recovery:**

1. Elevator -- RELEASE BACK PRESSURE.
2. Airspeed -- ACCELERATE to $V_Y$.
3. Power -- REDUCE to cruise RPM at desired altitude.

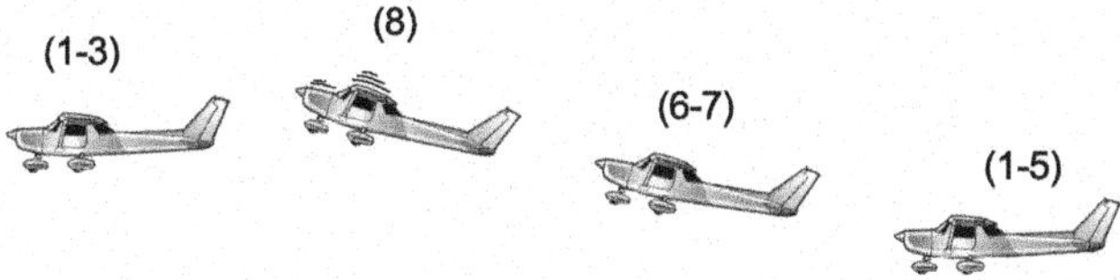


# TECHNIQUES

- Practice level and turning stalls.
- Use level pitch attitude for initial recovery and minimum loss of altitude. Allow airspeed to increase before re-establishing climb attitude to avoid secondary stalls.
- Use opposite rudder to correct a dropping wing.
- Re-establish a climb as soon as possible.

# APPROACH TO LANDING STALLS

**Objective:** To recover from a stalled condition in the landing configuration with a minimum loss of altitude and establish a positive rate of climb as soon as possible.

1.  Area -- CLEAR.
2.  Carburetor Heat -- AS REQUIRED.
3.  Power -- REDUCE to _______ RPM.
4.  Airspeed -- SLOW to final approach speed.
5.  Flaps -- FULL.
6.  Power -- CLOSE THROTTLE.
7.  Elevator -- INCREASE BACK PRESSURE until stall occurs.

## Recovery:

1.  Elevator -- RELEASE BACK PRESSURE.
2.  Power -- FULL THROTTLE.
3.  Carburetor Heat -- OFF.
4.  Altitude -- STABILIZE. Establish a climb.
5.  Flaps -- RETRACT TO _______ °.
6.  Airspeed -- _______ KTS. Accelerate to $V_Y$ while climbing.
7.  Flaps -- RETRACT SLOWLY while climbing.

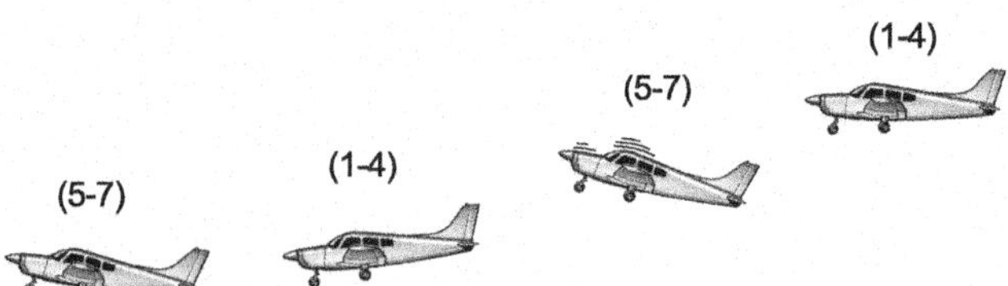


## TECHNIQUES

- Use a brisk forward elevator movement to break the stall.
- Avoid excessive nose down attitudes. Use level pitch attitude and full power to stop the descent. Note the initial upward movement of the vertical speed indicator.
- Retract the flaps one setting and establish a positive climb.
- Retract the remaining flaps while slowly accelerating to $V_Y$.

# SLOW FLIGHT/MINIMUM CONTROLLABLE AIRSPEED

1.  Area -- CLEAR.
2.  Carburetor Heat -- AS REQUIRED.
3.  Power -- REDUCE to ________ RPM.
4.  Altitude -- HOLD.
5.  Flaps -- FULL.
6.  Airspeed -- ________ KTS.
7.  Carburetor Heat -- OFF.
8.  Throttle -- MAINTAIN ALTITUDE ________ RPM.
9.  Trim -- ADJUST.
10. Rudder -- RIGHT RUDDER PRESSURE to maintain heading.

## Recovery:

1.  Power -- FULL THROTTLE.
2.  Flaps -- RETRACT.
3.  Airspeed -- ACCELERATE to normal cruise.
4.  Power -- REDUCE to ________ RPM.
5.  Trim -- ADJUST.

# TECHNIQUES

- Use shallow bank turns.
- Monitor the oil temperature while flying at slow airspeeds.
- Use right rudder pressure to maintain heading.
- Maintain vigilance for other aircraft.
- Remember: *Pitch for airspeed - Power for altitude.*

# ACCELERATED STALLS

1.  Area -- CLEAR.
2.  Carburetor Heat -- AS REQUIRED.
3.  Power -- REDUCE to _______ RPM.
4.  Airspeed -- _______ KTS.
5.  Bank -- ROLL INTO A 45º BANK.
6.  Elevator -- INCREASE BACK PRESSURE until stall occurs.

## Recovery:

1.  Elevator -- RELEASE BACK PRESSURE.
2.  Power -- FULL THROTTLE.
3.  Carburetor Heat -- OFF.
4.  Bank -- LEVEL WINGS.

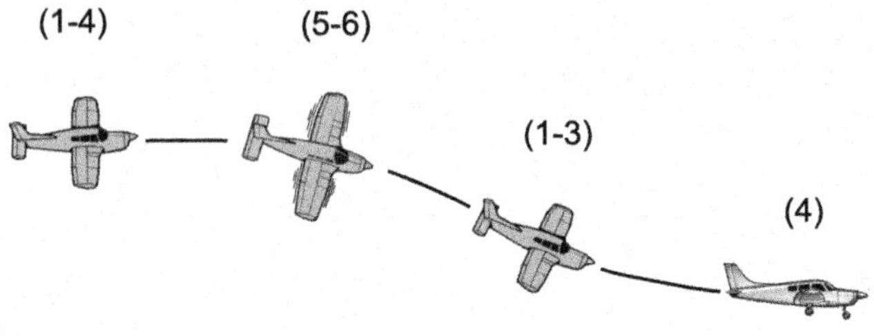


## TECHNIQUES

- Avoid abrupt control movements.
- Perform the maneuver only when operating below $V_A$.
- Slow to approach airspeed before turning.
- Reduce power and use positive continuous elevator back pressure once the turn has been established.

# STEEP TURNS 45° BANK

**Objective:** To maintain a constant altitude throughout a 360° turn in a 40° to 50° bank.

1. Area -- CLEAR.
2. Heading, Altitude -- STABILIZE.
3. Bank -- 45° BANK. Roll in promptly using coordinated aileron and rudder controls.
4. Elevator -- INCREASE BACK PRESSURE to hold altitude.
5. Power -- INCREASE slightly if required.
6. Altitude -- MAINTAIN using a reference point aligned with the horizon. Cross-check the flight instruments.
7. Bank -- MAINTAIN 45° bank. Use aileron controls to correct for overbanking tendency.
8. Rudder -- CENTERED BALL.
9. Roll-out -- 20° BEFORE desired heading roll out using positive coordinated rudder and aileron controls.
10. Level Flight -- REDUCE elevator back pressure and decrease power setting to maintain level flight.

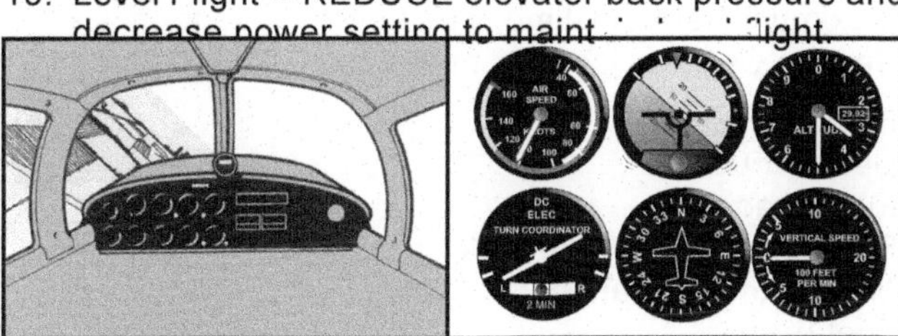

## TECHNIQUES

- Decrease the bank angle slightly before correcting for a low pitch attitude. Then resume a 45° bank.
- Use the dot on the horizon bar of the attitude indicator for a pitch reference in the absence of a good visual horizon.
- Remember from the pilots seat the nose will appear to rise during a left turn and drop during a right turn.

**MAKE AND MODEL** _______________________________

$V_{SO}$ - Stall Speed, Flaps Down, Power Off _______

$V_S$   - Stall Speed, Flaps Up, Power Off . . . . . . _______

$V_X$   - Best Angle of Climb Speed. . . . . . . . . . _______

$V_Y$   - Best Rate of Climb Speed . . . . . . . . . . _______

$V_{FE}$ - Maximum Flaps Extended Speed . . . . . . _______

$V_A$   - Maneuvering Speed. . . . . . . . . . . . . . _______

$V_{NO}$ - Maximum Structural Cruising Speed. . . . . _______

$V_{NE}$ - Never Exceed Speed   . . . . . . . . . . . _______

$V_R$   - Normal Takeoff Rotation Speed . . . . . . _______

Normal Climb Speed . . . . . . . . . . . . . . . . . . _______

Short Field Takeoff Climb Speed . . . . . . . . . . _______

Normal Approach Speed, Full Flaps . . . . . . . . . _______

Short Field Approach Speed . . . . . . . . . . . . . _______

Best Glide Speed. . . . . . . . . . . . . . . . . . . . _______

Maximum Crosswind Component. . . . . . . . . . . . _______

Maximum Rate of Climb, Sea Level . . . . . . . . . _______

Service Ceiling . . . . . . . . . . . . . . . . . . . . . _______

Fuel Capacity . . . . . . . . . . . . . . . . . . . . . . _______

Usable Fuel. . . . . . . . . . . . . . . . . . . . . . . . _______

Fuel Grade . . . . . . . . . . . . . . . . . . . . . . . . _______

Fuel Selector .. Off _____ Both _____ Left _____ Right _______

Maximum Endurance . . . . . . . . . . . . . . . . . . _______

Fuel Burn . . . . . . . . . . . . . . Gallons Per Hour _______

Oil Capacity. . . . . . . . . . . . . . . . . . . . . . . . _______

Minimum Safe Oil . . . . . . . . . . . . . . . . . . . . _______

Grade Oil. . . . . . . . . . . . . . . . . . . . . . . . . . _______

Maximum Weight . . . . . . . . . . . . . . . . . . . . _______

Empty Weight . . . . . . . . . . . . . . . . . . . . . . _______

Useful Load . . . . . . . . . . . . . . . . . . . . . . . . _______

Maximum Baggage . . . . . . . . . . . . . . . . . . . _______

Tire Pressures . . . . . . . . . . Nose _______ Main _______

# DENSITY ALTITUDE CHART

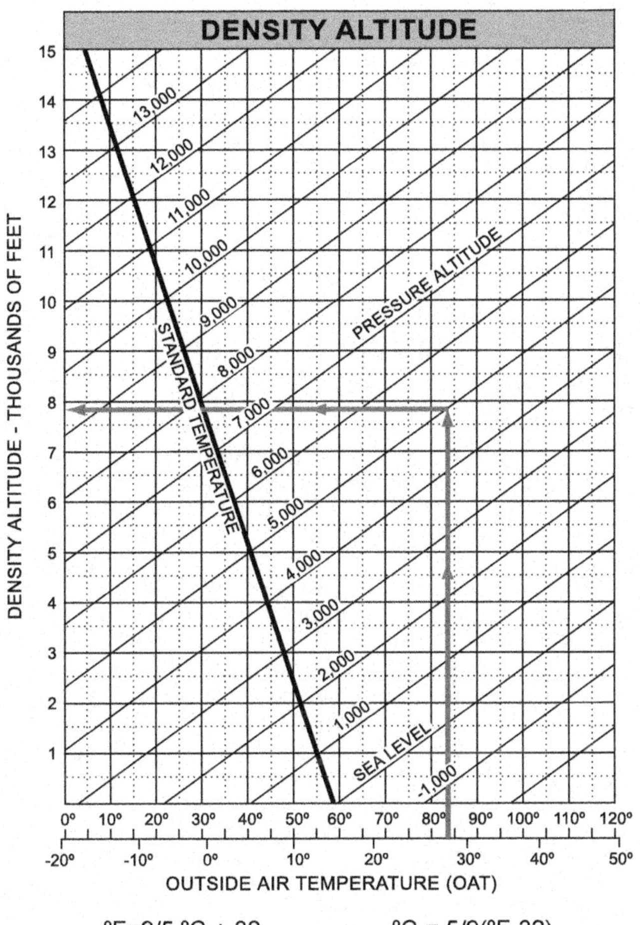

# TAKEOFF/LANDING CONSIDERATIONS BASED ON RUNWAY SLOPE AND WIND CONDITIONS

**Takeoff:**
- Headwind - Shorter takeoff roll.
- Crosswind Consider.
- Obstacles Consider.
- **Never trade a headwind for down hill slope.**

**Landing:**
- Into headwind - Avoid tailwind.
- Crosswind consider.
- Runway upslope - stop shorter / downhill longer.
- **Memory Aid: Wind / Slope / Go-Around Early.**
- **5 knots tailwind = 5% to 25% landing distance.**

## TAKEOFF AND LANDING SAFETY TIPS

1. Plan to touch down in the first third of the runway or go around.
2. Takeoff performance could determine a decision to land at a high altitude airport.
3. When possible fly a traffic pattern close enough to safely glide to the runway.
4. Over fly a runway in question. Look for holes or debris Note wind drift and ground speed.
5. Climb at best rate of climb to 500 feet AGL after takeoff.
6. Night landings. Fly a wide normal pattern. Avoid high airspeeds and low shallow final approaches. *Pilot Controlled Lighting - 7 clicks high intensity, 5 clicks medium intensity, 3 clicks low intensity.*
7. Remember not to attempt a turn back to the runway in the event of power failure after takeoff. *Note that maneuvering to land on the runway you just departed from requires in excess of 180° to as much as 270° of turning!*

# WEIGHT AND BALANCE COMPUTATION

1. Add weight of aircraft, passengers, baggage, and fuel.
2. Determine total weight to be within limits.
3. Multiply weight by arm for moment.
4. Add individual moments for total moment.
5. Divide the total moment by the total weight to determine the loaded center of gravity. (CG.)
6. Determine the loaded CG. to be within limits.

| MAX GROSS WT | LOADED C.G. | FWD LIMIT | AFT LIMIT |
|---|---|---|---|
|  |  |  |  |
|  | WEIGHT X ARM = MOMENT | | |
|  | (POUNDS) | (INCHES) | (INCH/ POUNDS) |
| AIRCRAFT BASIC EMPTY WEIGHT |  |  |  |
| PILOT SEAT AND FRONT PASSENGER |  |  |  |
| REAR SEAT PASSENGERS |  |  |  |
| FUEL GALLONS |  |  |  |
| BAGGAGE AREA |  |  |  |
|  |  |  |  |
| Total Moment / Total Weight = C.G. | Total Weight | Loaded C. G. | Total Moment |
|  |  |  |  |

Fuel = 6 lbs. per gallon
Oil = 1.875 lbs. per quart
Water = 8.3 lbs. per gallon

# AIRCRAFT LOADING

**Aft Loaded Aircraft** - Unstable pitch, poor stall recovery, higher cruise speed.

**Extreme Aft C.G.** - Very unstable, unable to recover from stall/spin condition.

**Forward Loaded Aircraft** - Stable pitch, lower cruise speed (more elevator down wash required).

**Extreme Forward CG.** - Unable to flare during landing, nose wheel strikes ground before main wheels.

**Over Maximum Gross Weight** - Structural failure in-flight if strong turbulence or excessive load factors are encountered, longer takeoff and landing distances, reduced climb performance, higher stalling speeds.

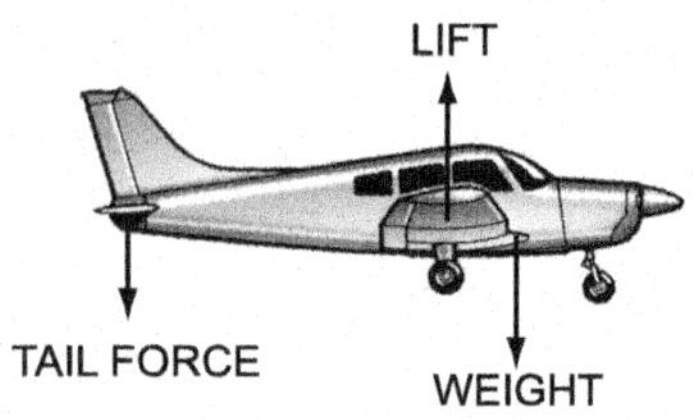


**Aircraft Empty Weight** - Includes the empty weight of the aircraft, undrainable fuel and hydraulic fluids, all installed equipment and engine oil (unless stated otherwise).

**Maximum Gross Weight** - The maximum certified loaded weight of the aircraft including fuel passengers and baggage.

**Useful Load** - Includes passengers, usable fuel, and baggage. The difference between the empty weight and the gross weight of the airplane.

# EMERGENCY PROCEDURES

## ENGINE FAILURE IMMEDIATELY AFTER TAKEOFF

1. Do not try to return to the airport!
2. Airspeed -- LOWER NOSE, maintain gliding airspeed.
3. Land -- STRAIGHT AHEAD minimal.
4. Mixture -- IDLE CUT-OFF.
5. Fuel Selector Valve -- OFF.
6. Ignition Switch -- OFF.
7. Flaps -- AS REQUIRED.
8. Master Switch -- OFF.

## ENGINE FAILURE IN FLIGHT

1. Airspeed -- BEST GLIDE _________ KTS.
2. Trim -- ADJUST.
3. Landing Area -- SELECT.
4. Restart: (if time and altitude permit)
5. Carburetor Heat -- ON.
6. Electric Fuel Pump -- AS REQUIRED.
7. Mixture -- FULL RICH.
8. Primer -- IN and LOCKED.
9. Fuel Selector Valve -- ON, BOTH or PROPER TANK.
10. Magneto Switch -- BOTH. Try left or right positions.
11. If unable to restart follow forced landing procedure.

## FORCED LANDING PROCEDURE

1. Airspeed -- BEST GLIDE _________ KTS.
2. Mixture -- IDLE CUT-OFF.
3. Fuel Selector Valve -- OFF.
4. Magneto Switch -- OFF.
5. Flaps -- FULL FLAPS ON FINAL APPROACH.
6. Master Switch -- OFF.
7. Doors -- UNLATCH BEFORE TOUCHDOWN.
8. Touchdown -- AS REQUIRED.
9. Brakes -- APPLY. Minimum ground roll.

**High Pitch Attitude Approaching a Stall -** Level pitch attitude, apply full power, level the wings with the horizon.

**Low Pitch Attitude Increasing Airspeed** - Level the wings with the horizon, reduce the power, level pitch attitude.

## SPIN RECOVERY

1. Ailerons -- NEUTRALIZE.
2. Throttle -- CLOSE.
3. Rudder -- HOLD FULL RUDDER opposite the direction of rotation.
4. Elevator -- MOVE QUICKLY FORWARD to break the stall.
5. Rudder -- NEUTRALIZE after rotation stops.
6. Elevator -- SMOOTHLY RECOVER from resulting dive.

## 180° TURN OUT OF CLOUDS

1. Remain calm. Fly the plane using the attitude indicator and the other flight instruments.
2. Note your present heading and determine the reciprocal heading.
3. Begin a standard rate turn using the turn coordinator. This will equal about a 15° to 20° bank on the attitude indicator.
4. After 1 minute roll out on the reciprocal heading.
5. Be patient. It may take several minutes to fly out of the clouds.

## EMERGENCY COMMUNICATIONS

*Squawk **7700**. Transmit on **121.5 MHz** if unable to reach a nearby ATC facility. Give the following information:*

1. **"MAYDAY"** (Repeated 3 times).
2. Name of the station you are calling.
3. Aircraft type and identification.
4. Nature of emergency or situation.
5. Position, heading, and altitude.
6. Fuel remaining on board.
7. Number of persons on board.
8. Weather conditions.
9. Your intentions or request.

### AVIATION ROUTINE WEATHER REPORT (METAR)

Scheduled hourly observation. **SPECI** indicates special report.

**METAR KPIT 201955Z 22015G25KT 3/4SM  28R/2600FT TSRA  OVC010CB  18/16  A2992  RMK RETSB23**

**Decoded Report** - Pittsburgh International Airport, **20th** day of the month, **1955 UTC**, wind **220** at **15** knots, gusting to **25** knots, visibility **3/4 statute mile**, runway **28R RVR is 2600** feet, thunderstorms and moderate **rain**, **1,000** foot overcast clouds consisting of **cumulonimbus** clouds, temperature **18°** Celsius, dewpoint **16°** Celsius, altimeter setting **29.92"**. Re**marks**: recent weather event thunderstorm began **23** minutes past the hour.

## WEATHER DECODER

| | | | | | |
|---|---|---|---|---|---|
| **GS** | Snow Pellets | **TS** | Thunder Showers | **UP** | Unknown Precipitation |
| **BL** | Blowing | **MI** | Shallow | **VA** | Volcanic Ash |
| **BR** | Mist | **PE** | Pellets | **SKC** | Sky Clear |
| **DR** | Low Drifting | **PO** | Dust/ Sand Whirls | **CLR** | Clear |
| **DU** | Dust | **PR** | Partial | **FEW** | 1/8 - 2/8 |
| **DS** | Dust Storm | **PY** | Spray | **SCT** | 3/8 - 4/8 |
| **DZ** | Drizzle | **RA** | Rain | **BKN** | 5/8 - 7/8 |
| **FG** | Fog | **SA** | Sand | **OVC** | 8/8 |
| **FC** | Funnel Cloud | **TCU** | Towering Cumulus | **VV** | Vertical Visibility |
| **FU** | Smoke | **SH** | Showers | **CB** | Cumulonimbus |
| **FZ** | Freezing | **SN** | Snow | **SG** | Snow Grains |
| **GR** | Hail | **SQ** | Squall | | - Light |
| **BC** | Patches | **SS** | Sand Storm | | Moderate (no sign) |
| **HZ** | Haze | **IC** | Ice Crystals | | + Heavy |

Detailed 24 or 30 hour forecast of the weather within a 5 NM radius of the airport. May include expected weather within 5NM to 10 NM (preceded by **VC** for **Vic**inity).

```
TAF KPIT 091730Z 0918/1024 15005KT 5SM HZ
      FEW020 WS010/31022KT
      FM 091930 30015G25KT 3SM SHRA OVC015
      TEMPO 0920/0922 1/2SM +TSRA OVC008CB
      FM100100 27008KT 5SM SHRA BKN020 OVC040
      PROB30 1004/1007 1SM -RA BR
      FM101015 18005KT 6SM -SHRA OVC020
      BECOMG 1013/1015 P6SM NSW SKC
```

**Decoded forecast:** Pittsburgh, PA **(KPIT)** terminal forecast issued on the **9**th at 5:30 pm UTC (**1730Z**) valid from the **9**th at 6:00 pm UTC (**1800Z**) through the **10**th at 12 midnight UTC (**2400Z**), wind **150°** at **5** knots, visibility **5 s**tatute **m**iles, **haze, few** clouds at **2**,000 feet.

**WS010/31022KT** means low level wind shear expected at 1,000 feet, wind 310 degrees at 22 knots.

**FM091930** means on the **9**th **f**rom 7:30 pm UTC (**1930**Z)

**TEMPO 0920/0922** means **tempo**rary conditions expected on the **9**th between 8:00 pm UTC (**2000Z**) and 10:00 pm UTC (**2200Z**)

**FM100100** means on **10**th day from 1:00 am UTC (**0100Z**)

**PROB30 1004/1007** means there is a **30% prob**ability of the conditions on the **10**th day between **04**00Z and **07**00Z

**BECOMG** means conditions becoming as described on the **10**th day between **13**00Z and **15**00Z



Forecast of general weather conditions over an area the size of several states. Contains a 12-hour forecast with a 6 hour outlook issued 3 times per day. Used for enroute weather and to interpolate forecasts for airports that do not have a **TAF**.

1. What altitude must be maintained to over-fly Whiteman Airport (WHP) and not be in communication with the control tower?
2. How high are the obstructions to the northeast of Santa Monica Airport?
3. What frequency would you self announce your positions when landing at Van Nuys Airport after the tower was closed?
4. What is the floor of the Class B airspace overlying overlying Dodger Stadium and downtown Los Angeles?
5. What does the dashed line circle and square extension surrounding Santa Monica Airport identify?
6. Flying westbound along V597 at 4,500' what are the requirements to operate in that airspace?
7. What is the elevation of Van Nuys Airport?
8. How long is the longest runway at Los Angeles Airport?
9. What does the *L 49 at Hawthorne airport indicate?
10. Can you fly directly over LAX below 10,000' without an ATC clearance?
11. On what frequency would you contact SOCAL Approach to enter the Class C airspace?
12. Is a Mode C transponder and ADS-B out required to take off from Van Nuys Airport? At 7,000' MSL above Van Nuys Airport? At 12,700' MSL above LAX?
13. Why is the frequency for Los Angeles VOR underlined?
14. How would you identify the Van Nuys VOR?
15. How would you locate Compton Airport at Night?
16. Why is Hawthorne Airport blue and Compton Airport magenta?
17. What does RP 7L, 7R indicate for the Compton airport?

*(Answers on page 48)*

NOT TO BE USED FOR NAVIGATION

(47)

# SECTIONAL CHART REVIEW

1. Above 3,000' MSL over Whiteman Airport.
2. Twin High Rise 866' msl. UCLA buildings 667' msl.
3. 119.3 as indicated by the © following the frequency.
4. 2500' MSL over downtown buildings and Doger Stadium.
5. Class D airspace. Special VFR is required in this airspace when conditions are less than 1,000' ceiling or 3 miles visibility to separate VFR and IFR traffic.
6. Magenta Circle is Class C airspace. Contact with ATC is required and have an operational Mode C transponder and ADS-B equipment.
7. 802' above mean sea level.
8. 12,900 feet long.
9. *L indicates pilot controlled lighting sunset to sunrise, consult AFD for details, 49 indicates the longest runway to be 4900 feet long.
10. Only in the special flight rules area, at the correct altitude for direction of flight, and requires the possession of a Los Angeles Terminal Area Chart.
11. 124.6 as indicated in the magenta box.
12. Yes. Van Nuys is within a 30 mile radius of the primary airport in a Class B airspace (LAX). Yes, over-flying Class C airspace. Yes Mode C is required at all times above 10,000' MSL.
13. No voice transmission is available.
14. The Morse code is depicted in the VOR box.
15. The airport has a rotating beacon as indicated by the star with a circle in it next to the airport symbol. Since Compton is a civilian airport the pilot would see alternating green and white flashes.
16. Airports with control towers are printed in blue. Uncontrolled airports are printed in magenta.
17. RP 7L,7R indicates a right hand traffic pattern for Runway 7 Left and Runway 7 Right.

## FLIGHT SERVICE STATIONS

**Aircraft:** *"Hawthorne Radio, Cherokee 15831 on 122.35."*
**FSS:** *"Cherokee 15831, go ahead."*
**Aircraft:** *"Cherokee 15831 departed Van Nuys airport at 10:45 local time please activate our VFR flight plan to Las Vegas."*

## ENROUTE FLIGHT ADVISORY SERVICE (EFAS)

Enroute flight advisories are provided by FSS for obtaining weather updates and giving pilot reports.

**Aircraft:** *"Cessna 5229D calling Hawthorne Flight Service"*
**FSS:** *"Cessna 5229D, Hawthorne Radio, go ahead."*
**Aircraft:** *"Cessna 5229D at 5,500' over Palmdale VOR, enroute to Las Vegas, request current weather at McCarran airport."*

## PILOT REPORT

Each pilot report should include the aircraft type, location, altitude, and one or more of the following items:

1) Clouds: Height of bases or tops, IFR conditions
2) Visibility: In-flight visibility, ground fog, blowing dust.
3) Temperature: Outside air temperature.
4) Winds Aloft: Strong headwinds or tailwinds.
5) Turbulence: Smooth, light, moderate or severe.
6) Icing: Carburetor icing or icing in clouds.
7) Remarks: Dark nights, bird flocks, forest fires, etc.

**Aircraft:** *"Hawthorne Flight Service, N1549L is a Grumman Tiger at 4,500' over CMA VOR with a pilot report."*
**FSS:** *"Grumman 1549L go ahead with your report."*
**Aircraft:** *"Grumman 1549L departed Van Nuys at 10:00 local, smooth with occasional light turbulence during climb. Base of the overcast is at 6,000'."*

# RADAR SERVICE (FLIGHT FOLLOWING)

Radar service frequencies can be found on the Sectional Chart frequency tabs, Airport Facilities Directory, Airport Guides, or IFR Enroute Charts.

**Aircraft:** *"LA Center, Apache 4164P, request."*
**ATC:** *"Apache 4164P, LA Center, go ahead."*
**Aircraft:** *"Apache 4164P, is a PA-23/A over Filmore VOR at 6,500 VFR enroute to Santa Barbara, request VFR Flight Following."*
**ATC:** *"Apache 4164P squawk 4322 and ident."*
**ATC:** *"Apache 4164P radar contact 2 miles west of the Filmore VOR, traffic 12 o'clock, 4 miles opposite direction altitude indicates 5,500."*
**Aircraft:** *"Six four Papa looking for traffic."*

## CLASS B AIRSPACE

An ATC clearance is required before entering Class B airspace. Listen for the controller to say ***"Cleared into the Class B airspace"***.

**Aircraft:** *"Las Vegas approach, Piper 3025N request."*
**ATC:** *"Piper 3025N Las Vegas approach, go ahead."*
**Aircraft:** *"Piper 3025N is a PA32/A, 30 miles south-west of Las Vegas VOR, at 9,500, landing at McCarran with ATIS information Charlie."*
**ATC:** *"Piper 3025N remain clear of the Class Bravo airspace, squawk 4663, Las Vegas altimeter setting 30.02."*
**Aircraft:** *"30.02, 25 November."*
**ATC:** *"Piper 3025N radar contact 27 miles south-west of McCarran, cleared into the Class B airspace fly heading 040° descend to and maintain 4,000."*
**Aircraft:** *"Heading 040°, descend and maintain 4,000 for Piper 25 November."*
**ATC:** *"Piper 3025N contact Las Vegas approach on 119.4."*
**Aircraft:** *"One one niner point four, Piper 3025 November."*

## CLASS C AIRSPACE

Class C airspace requires two-way radio communications before entering the airspace which is normally established when the controller acknowledges your N-number.

**Aircraft:** *"So-Cal approach, Navajo 4200N, request."*

**ATC:** *"Navajo 4200N, So-Cal approach, go ahead."*

**Aircraft:** *"Navajo 4200N 2 miles south of Van Nuys airport climbing through 1,700' to 7,500, eastbound to El Monte airport."*

**ATC:** *"Navajo 4200N squawk 5374 Burbank altimeter 29.98."*

## GROUND CONTROL

All communications with ground control should include:

- **WHO** - *you are:*    Aircraft type and N-number.
- **WHERE** - *you are:*    The location on the airport.
- **WHAT** - *you want:*    Where you wish to taxi to.

**Aircraft:** *"Santa Monica ground, Cessna 8306X, at transient parking, taxi for take-off with Charlie".*

## SPECIAL VFR

### On the Ground:

**Aircraft:** *"Santa Monica Ground Mooney 5229H, transient parking with ATIS information Bravo, request special VFR westbound to Oxnard airport."*

**Ground:** *"Mooney 5229H cleared special VFR westbound, remain clear of clouds, report clear of the Class D airspace."*

### In-flight:

**Aircraft:** *"Santa Monica tower, Cessna 8306X , over UCLA landing Santa Monica with X-ray, request special VFR."*

**ATC:** *"Cessna 8306X, squawk 5534, cleared into the Class D airspace, enter right base over the freeway, report a 2 mile final."*

## UNCONTROLLED AIRPORT

At airports without an operating control tower pilots should self-announce their position and intentions on the **Common Traffic Advisory Frequency (CTAF).**

**Aircraft:** *Santa Paula traffic Cherokee 15831 is 10 miles east of the airport request airport information."*

*"Santa Paula traffic Cherokee 15831 entering left downwind, runway 22 Santa Paula."*

*"Santa Paula traffic Cherokee 15831 left base, runway 22 Santa Paula."*

*"Santa Paula traffic Cherokee 15831 final, runway 22 Santa Paula."*

*"Santa Paula traffic, Experimental N106NC departing runway 22, left downwind departure."*

## CONTROL TOWER (CLASS D AIRSPACE)

**Aircraft:** *"Santa Monica tower, Apache 4162P at runway 21, ready for takeoff."*
**Tower:** *"Apache 4162P hold short for landing traffic."*
**Aircraft:** *"Six two Papa, holding short." (Always repeat back all hold short instructions).*

## CLEARANCE DELIVERY

Listen to the ATIS. Contact Clearance Delivery, state your N-number, location, destination and VFR cruising altitude".

**Aircraft:** *"Las Vegas Clearance, Cessna 7396L at Atlantic Aviation with ATIS information Zulu, VFR to Van Nuys at 8,500."*
**ATC:** *"Cessna 7396L, turn left heading 170° after departure, climb to and maintain 3,500, expect higher altitude in 2 minutes, departure control is 118.4, squawk 5536."*
**Readback:** *"Left turn to 170°, climb to 3,500', higher in 2 minutes, 118.4 squawk 5536 for 96L."*

# FLIGHT PLANNING

## CROSS-COUNTRY CHECKLIST

1.   Sectional Charts.
2.   Aircraft Flight Manual.
3.   Airport Facilities Directory or Airport Guide.
4.   Clipboard and Pencils.
5.   Flight Computer and Plotter.
6.   Airsick Bags.
7.   Warm Clothing.
8.   Flashlight(s).
9.   Headsets or extra Microphone.
10.  Fuel Tester.
11.  Fire Extinguisher.
12.  Survival Equipment and Water.
13.  Life Jackets (over water).
14.  Pilot's Certificate and Medical Certificate.

## AIRCRAFT CHECKLIST

1.   **A.R.R.O.W.**
2.   Aircraft Lights -- OPERATIONAL.
3.   Engine Oil -- FULL. Extra quarts if needed.
4.   Tires -- PROPER INFLATION.
5.   Windshield -- CLEAN.
6.   Instrument Panel Lights -- OPERATIONAL.

## ROUTE PLANNING CHECKLIST

1.   Aircraft Takeoff and Landing Performance.
2.   Fuel Available at Destination and Alternatives.
3.   Weather Briefing.
4.   Flight Plan Filed. Departure and Destination Contacts
     should know; Type Aircraft, N-number, Airport Name,
     FBO, Persons on Board, Telephone Numbers.
5.   ATC Radar and FSS frequencies along the route.
6.   Altitude/Oxygen Requirements (8,000' night).
7.   Flight Plan Alternatives.

## WEATHER BRIEFING FORMAT

1. Adverse Conditions.
2. Synopsis.
3. Current Weather including PIREPS.
4. Enroute/Destination Forecast.
5. Winds and Temperatures Aloft.
6. Notams.

## ICAO FLIGHT PLAN

- Aircraft Identification
- Flight Rules (**V** for **VFR. I** for **IFR**)
- Type of Flight (**G** for General Aviation)
- Number of aircraft (**1** unless formation flgiht)
- Type of Aircraft (**P28A, C172**, etc)
- Aircraft Equipment (Nav/Com, Transponder, ADS-B)
- Departure Airport,
- Estimated Time of Departure UTC (ETD)
- Cruising Speed (**N** for knots, followed by 4 digits)
- Level (**A** for Altitude, followed by 3 digits)
- Route of Flight (Waypoints, Airways and VOR's)
- Destination Airport
- Estimated Enroute Time (EET)
- Other Information (RMK for remarks)
- Suplementary Information:
  (**E/** Fuel Endurance Hours and MInutes)
  (**P/** Persons on Board)
  (**A/** Aircraft Color, Markings)
  (**C/** Pilot Contact Information)

| SPECIAL EQUIPMENT SUFFIXES | |
| --- | --- |
| /X | No Transponder |
| /T | Transponder – no Mode C |
| /U | Transponder – Mode C |
| /B | DME-Transponder – no Mode C |
| /A | DME-Transponder - Mode C |
| /V | GPS – No Transponder |
| /S | GPS – Transponder - no Mode C |
| /G | GPS – Transponder - Mode C |

## VFR GPS CHECKLIST

(Before Engine Start or Before Taxi)

### POWER & STATUS

- ☐ GPS ON / booted
- ☐ Position acquired (no alerts)
- ☐ Database current (or knowingly expired for VFR)
- ☐ Correct aircraft profile (if applicable)

### ROUTE

- ☐ Departure airport entered
- ☐ Destination airport entered
- ☐ Route matches planned VFR course
- ☐ First leg points the correct direction

### VERIFY (DO NOT SKIP)

- ☐ Scroll all legs
- ☐ Waypoint names make geographic sense
- ☐ Distances reasonable
- ☐ Magenta line matches chart / eyeball plan

Say out loud: "From _______ to _______ via _______."

### NAV MODE

- ☐ GPS mode selected (not VLOC)
- ☐ CDI centered & responding
- ☐ OBS off (unless intentionally used)

### AIRSPACE & TERRAIN

- ☐ Airspace alerts ON
- ☐ Terrain alerts ON (if available)
- ☐ Class B/C/D reviewed along route

### BACKUP

- ☐ EFB or paper chart available
- ☐ Nearest function reviewed
- ☐ Lost-GPS plan briefed (heading + time)

### FINAL CHECK

- ☐ First turn expected? YES / NO
- ☐ GPS is aid only — not primary VFR nav

# NAVIGATION LOG

| A/C NUMBER | A/C TYPE | DEPARTURE | DESTINATION | DATE |
|---|---|---|---|---|
|  |  |  |  |  |

| Departure Point | V O R Ident. / Freq. | Radial To / From | Altitude | Route Magnetic Course | Distance Leg / Total | Ground Speed | Time Leg / Total |
|---|---|---|---|---|---|---|---|
|  |  |  |  |  |  |  |  |
|  |  |  |  |  |  |  |  |
|  |  |  |  |  |  |  |  |
|  |  |  |  |  |  |  |  |
|  |  |  |  |  |  |  |  |
|  |  |  |  |  |  |  |  |
|  |  |  |  |  |  |  |  |
| Destination |  |  |  |  |  |  |  |

## ICAO FLIGHT PLAN

| AIRCRAFT ID | FLIGHT RULE | FLIGHT TYPE | NUMBER OF AIRCRAFT | AIRCRAFT TYPE | WAKE TURBULENCE |
|---|---|---|---|---|---|
|  |  |  |  |  |  |

| AIRCRAFT EQUIPMENT | SURVEILLANCE EQUIPMENT | DEPARTURE | CRUISING SPEED | LEVEL |
|---|---|---|---|---|
|  |  |  |  |  |

| ROUTE | DESTINATION AIRPORT | ESTIMATED ELAPSED TIME |
|---|---|---|
|  |  |  |

| FUEL ENDURANCE | PERSONS ON BOARD | COLOR OF AIRCRAFT | PILOT IN COMMAND |
|---|---|---|---|
|  |  |  |  |

OTHER INFORMATION:

CLOSE VFR FLIGHT PLAN WITH __________ FSS ON ARRIVAL

**FLIGHT SERVICE**  FAA certified weather and flight planning. 1800wxbrief.com

**FltPlan.com** Flight Planning for General Aviation. Weather, airport information, flight tracking, approach plates. fltplan.com

**AOPA.org**  Aircraft Owners and Pilots Association. Flight planning, weather, training and safety courses. aopa.org

**FOREFLIGHT** FAA certified weather and flight planning. foreflight.com

**NOAA's National Weather Service Aviation Weather Center** aviationweather.gov

**PilotWeb** FAA NOTAM's website. GPS NOTAM's, RAIM predictions, TFR's. pilotweb.nas.faa.gov

**AIRNAV.COM** Airport search and information, NACO charts. airnav.com

**TFR.faa.gov** FAA listing of Temporary Flight Restrictions tfr.faa.gov

**Federal Aviation Administration** links to all government publications. Aeronautical Information Manual, FAR regulations, Airman Certification Standards, Manuals and Advisory Circulars (AC's) . faa.gov

**SkyVector.com** Sectional Charts, Airport Diagrams, Enroute Charts, METAR, TAF, AF/D.  skyvector.com

**Intellicast.com** WSI weather radar, local and global weather charts and forecasts, ipad apps. Member of The Weather Channel Group. intellicast.com